Elon Musk

Elon Musk's Best Lessons for Life, Business, Success and Entrepreneurship!

Table of Contents

Introduction 5

Chapter 1: Get to Know Elon Musk 6

Chapter 2: Growing Up 8

Chapter 3: Early Businesses 11

Chapter 4: Current Projects 15

Chapter 5: What's Next 24

Chapter 6: Solar City 27

Chapter 7: Elon's Investments 29

Chapter 8: Personal Life 37

Chapter 9: Artificial Intelligence 39

Chapter 10: What Will Be Elon Musk's Legacy? 41

Chapter 11: Lessons to Learn 43

Conclusion 47

This document, as presented, with the desire to provide reliable, quality information about the topic in question and the facts discussed within. This eBook is sold under the assumption that neither the author nor the publisher should be asked to provide the services discussed within. If any discussion, professional or legal, is otherwise required a proper professional should be consulted.

This Declaration was held acceptable and equally approved by the Committee of Publishers and Associations as well as the American Bar Association.

The reproduction, duplication or transmission of any of the included information is considered illegal whether done in print or electronically. Creating a recorded copy or a secondary copy of this work is also prohibited unless the action of doing so is first cleared through the Publisher and condoned in writing. All rights reserved.

Any information contained in the following pages is considered accurate and truthful and that any liability through inattention or by any use or misuse of the topics discussed within falls solely on the reader. There are no cases in which the Publisher of this work can be held responsible or be asked to provide reparations for any loss of monetary gain or other damages which may be caused by following the presented information in any way shape or form.

The following information is presented purely for informative purposes and is therefore considered universal. The information presented within is done so without a contract or any other type of assurance as to its quality or validity.

Any trademarks which are used are done so without consent and any use of the same does not imply consent or permission was gained from the owner. Any trademarks or brands found within are purely used for clarification purposes and no owners are in anyway affiliated with this work.

Introduction

Thanks and congratulations on picking up this book - *Elon Musk: Best Lessons for Life, Business, Success and Entrepreneurship!* This is the recently updated 2nd edition of this book, complete with several additional chapters and improvements!

When trying to improve your own lot in life it is always a great idea to look at the successful habits of those who came before and there are few better examples of success in the twenty first century than Elon Musk.

More than just a shrewd businessman, engineer or inventor, Musk is truly a visionary, someone who is working to improve the world not because it can make him money, or even because it is the right thing to do, but simply because he sees humanity's true potential and wants it to stick around long enough to come to fruition.

Inside you will find an account of Musk's life from his time as a young victim of bullying in South Africa to creating his first business, to being the CEO of a pair of companies worth more than a billion dollars each. There is plenty to learn from Musk's life and an in-depth discussion of those lessons can be found in the final chapter.

Thanks for purchasing this book, I hope it provides you with everything you are looking for. Enjoy!

Chapter 1:
Get to Know Elon Musk

While Elon is the first Musk that the people of the twenty first century have heard of, it isn't the first time one of their ranks has risen to distinguished prominence on the international stage. Elon's grandparents were the first to travel by airplane from Australia to Africa, his great grandfather won a well popularized race across the length of Africa and his other grandmother was the first woman to ever hold a Canadian chiropractor's license.

With a lineage like this, the Musk family has always thought of itself as pioneers, an affiliation that the latest, and possibly greatest, Musk has taken to heart throughout his adult life. So far, he has been involved in creating one of the most ubiquitous websites throughout the whole of the internet, gotten behind electric cars in a significant way and set his sights on interplanetary space travel within the next 20 years - and all before his fiftieth birthday!

Musk is also a symbol of the American dream, despite not being American, he is truly self-made, working his way through college and starting his first business in time to take advantage of the burgeoning internet scene to become a millionaire before the age of 30. Unlike many people who might think that was enough, Musk instead put his money into another startup which quintupled his net worth when his company was purchased by eBay.

Still not satisfied, Musk then sunk the entirety of what he had earned up until that point, into betting on the future and betting big. This is what ultimately lead him to the mainstream limelight as he took on a leading role at Tesla Motors and

SpaceX. He is also heavily invested in solar energy, owning a major share of one of the biggest solar power producers in California. He is also actively funding what is likely the future of long distance travel by helping to make what is known as the Hyperloop a reality.

Finally, when he finds a bit of spare time, he is working to ensure that what he feels is the inevitable creation of artificial intelligence doesn't end up working out in any way that doesn't benefit humanity as a whole. What it comes down to is that if there is a futuristic and implausible sounding technology that will make the world a better place down the line, then it is a safe bet that Musk is on the frontlines and has already committed more than a million dollars to ensure it becomes a reality.

While certainly of genius level intelligence, that isn't what makes Musk such a positive source for change; no, that honor goes to his ability to predict likely future outcomes based on present indicators and his ability quickly absorb vast amounts of new information and utilize it properly. This ability can be seen most recently through his SpaceX venture as when he became interested in actually sending something to Mars he started by reading everything he could about rocket science and talking to a bunch of rocket scientists. From that point on he could accurately predict where the market for such things was going because he comprehensively understood the science behind it.

With his unique blend of knowledge acquisition skills, coupled with a business acumen that has been proven time and time again, Musk could, without question, do just about anything he wants. Humanity as a whole is extremely fortunate that what he has decided on is making the world into the science fiction utopia scientists have been chasing for a hundred years.

Chapter 2:
Growing Up

The man who many people are calling the modern day Henry Ford was born to Maye and Errol Musk on June 28, 1971 in South Africa. The eldest of three children, Elon quickly took after his father, an electrical engineer, and while often considered introverted and quiet, by the age of 10 he was already programing in BASIC. He taught himself this skill on his father's Commodore VIC-20 and quickly used the skills he learned to start his first business.

By the age of 12 he had programed his own game, titled Blastar, and quickly sold it to a local magazine company who owned a publication dedicated to the emerging phenomenon for a total profit of $500. Despite his early entrepreneurial success, childhood was frequently difficult for the young Musk who was small for his age and was frequently bullied by the other children at the numerous private schools he attended. This bullying grew to be so severe that while in middle school he was actually thrown down a flight of stairs and beaten so badly that he was hospitalized after the attack.

By the time Musk was a teenager, the topic of Apartheid was hotly contested and he was looking for any way to avoid his mandatory service in the South African military who busied themselves during this period by actively silencing those speaking out against the practice. At the age of 17, and finished with secondary school, he originally tried to gain access to the United States and its burgeoning computer technology scene but was denied entry into the country. Not one to be stymied by an initial refusal, he instead went to Canada in 1989 using his mother's Canadian heritage to gain easy entrance into the country.

College

With his time in the military avoided, Musk had done little to actually improve his lot and he spent the next year living hand to mouth, saving every penny he could in order to afford tuition at a local college. He wasn't picky about the type of work he did either, working hard, menial jobs tending vegetables, shoveling grain, chopping wood and even cleaning the boiler at a local saw mill, a task that few stuck with for any length of time. When he was told this at the time, Musk wasn't sure what all the fuss was about; he found out quickly, however, as the job required a hazmat suit and for him to stay in a cramped, hot and dangerous space for several hours at a time. Despite the difficulty, Musk persevered and lasted significantly longer than most other people who attempted the task.

With his college tuition paid for, Musk was able to earn a spot at the Ontario-based Queens University. He spent 2 years attending the institution and spending time with his mother and younger brother Kimbal. In their spare time, he and Kimbal would read the newspaper and then call the interesting people they read about and try and have lunch with them. One such person they called was the president of a local bank, the resulting lunch landed Musk an internship and a staunch supporter in his future endeavors.

At this point Musk was just 18, though he was already looking towards the future and how he could make his mark on it. So much so, that the bank manager's daughter can still remember a conversation they had at Musk's eighteenth birthday. It was about electric cars.

By 1992, Musk had set himself apart from his peers by excelling at his studies, so much so that he was able to earn a scholarship to a college in the United States. He spent the next year studying at the University of Pennsylvania before earning a bachelor's degree in Physics before earning another the following year in Economics. Despite his apparent success, Musk found himself battling against depression during this period. He turned to the religious texts of various religions for guidance, before ultimately finding what he was looking for in a book by Douglas Adams called *The Hitchhiker's Guide to the Galaxy*.

While reading the book, he came across a passage which revealed to the main character that the meaning of life was 42. Where many people would see a simple joke, this answer stuck with Musk and taught him the importance of asking the correct question in any situation. With the answer to the meaning of life ringing in his mind, he thought about it for a while and then asked himself what he thought of as the right question.

This question was what technologies were likely going to have the greatest effect on the entire human race in the near future. His answers were the renewable energy, space travel and the then-burgeoning world wide web. With these ideas in mind, Musk applied, and was quickly accepted, into a doctorate program at Stanford University to study applied physics. Two days into his time at Stanford, he left the school to form his first company Zip2. Four years later he was worth more than a million dollars.

Chapter 3:
Early Businesses

In the months prior to his traveling to California, Musk watched in fascination as Netscape Communications went public and made a man younger than himself extremely wealthy in the process. At this point Musk's assets totaled out at approximately $2,000 and a used car but he knew what he wanted to do and how to go about doing it. He brought his brother Kimbal out to California, borrowed $28,000 from his father and started a software company named Zip2.

Zip2

Zip2 was a business that sold city guides that could then be used by newspapers as a way to make themselves relevant in the fledgling online space. Zip2 was created after Musk met the creators of Navteq, a digital mapping company and convinced them to let him use their online maps. He then purchased a digital directory of businesses in his area, combined them using a little bit of code and created one of the first digital mapping services.

Armed with his innovative product, Musk soon found himself courting numerous major newspapers and soon after that, offering his service via their websites. Despite their early success, the Zip2 offices were in a rundown office building that also served as the Musk brothers' primary place of residence. The ceiling leaked excessively and the only furniture to speak of was a pair of futons and a desk for Musk's computer which also served as the primary server for all of Zip2's traffic. During the evenings Musk would use the program to code and during the day it would connect to the internet via a hole in the floor Musk drilled to take advantage of the online connection available in the business downstairs.

As Zip2 grew, it began to attract the attention of investors and in 1996 a venture capital firm offered the company $ 3 million, but Musk had to step aside as the CEO and let a business executive named Richard Sorkin take over. The venture capital firm felt Sorkin was a better choice because he had a degree from Stanford. Initially it seemed as though everything was going to be smooth sailing as Sorkin quickly landed contracts with major newspapers such as the *Chicago Sun* as well as the *New York Times*. Unfortunately for Musk, Sorkin then began to seek out additional investment income from several of these large newspaper chains as well.

Musk didn't like the idea of taking on investors from the customer base the company was supposed to be targeting but found himself with little to do as by the time he realized what was happening he only owned 7 percent of all of Zip2's stock. As such, he was stuck watching other startups such as Yahoo! moving towards the future while his company was deeply entrenched in the ways of old media. Despite founding the company, he was now stuck in a vice president role with little to do but watch his company's relevance slip away.

Later that same year, Musk got word that Sorkin was working on a deal that would sell Zip2 to a search engine named CitySearch which would have created a nationwide version of the local search engine. Musk decided that this was where he had to draw the line or be left with nothing to fight for. He spread word of his dissention within the company and staged a revolt in an effort to get Sorkin removed from power. With many of the biggest names in the company behind him, Musk managed to successfully get Sorkin removed from the company.

This is where his success stymied, however, as instead of reinstating him as CEO as had been his plan, the board was still made up primarily of Newspapers and other members of the old guard who instead put Mohr Davidow's Derek Proudian in charge instead. As such, while the CitySearch deal was crushed, Zip2 was soon sold to Compaq instead who paid $300 million, the most an internet company had ever been valued at, and folded it into its conglomerate with no place in the package for Musk.

At the time, Musk was critical of the sale noting that the venture capital firm should have left him in charge as "Great things never happen with professional managers or VCs [venture capitalist firms] in charge, they don't have the insight or the creativity."

At its peak Zip2 served 120 newspaper groups countrywide, and its ultimate sale netted him a personal $22 million. Despite this, Musk has always considered Zip2 a failure because he had hoped to shape the internet more actively, with the benefit of the average person his primary concern. The fact that he instead created a way for a dying industry to remain relevant for a handful of additional years stuck with him which is why his next venture was designed from the ground up to challenge the status quo at every turn.

X.com

In 1999, Musk was ready to pitch his next idea, a financial services platform which worked without the need for traditional banks, to Sequoia Capital a well-known firm that the likes of Oracle, Cisco and Apple had all used to get their start. That same day he walked out of the meeting with $25 million in investment capital for his next venture, a website known as X.com.

While the original idea that Musk pitched to Sequoia included many if not all of the features that PayPal offers today, the version of X.com that launched in 1999 was significantly scaled back and instead hyper-focused on making person to person online payments a reality. The company saw early success and he was soon offered an attractive deal from a company called Confinity. Confinity wanted to merge with X.com but leave Musk in charge as the CEO of the new company. More importantly, it would provide Musk access to Confinity's PayPal software which was directly competing with X.com.

While the technology that each company brought to the table blended quite well, the same could not be said of the individuals that worked on the previously separate X.com and PayPal teams. The hostility and micromanaging meant that Musk spent the better part of a year competing with other personalities, visions and egos. This wasn't what Confinity founders Peter Thiel and Max Levchin were looking for when they agreed to let Musk be CEO, however, and the next year when Musk left the country to meet with new investors Levchin and Thiel used his absence to dispose him from his position as CEO.

The control of the company reverted to the Confinity pair who changed the name of the company to PayPal and sold it to eBay the next year for $1.5 billion in cash and shares of eBay stock. While no longer in control of yet another company, Musk still held 11.7 percent of PayPal's total stock at this juncture which meant the sale left him with $160 million.

Chapter 4:
Current Projects

While he certainly has numerous other pots in the fire, Musk is currently primarily focused on running two businesses, SpaceX and Tesla Motors and both are actively changing preconceived notions about their respective industries.

SpaceX

Musk has never been shy about sharing his opinions on space travel, which he believes is crucial to the long term survival of humankind. He is also fond of saying that he hopes he is going to die on Mars, but not during the initial landing. With that in mind, is it any wonder that he spent the time after he was removed as CEO of the future PayPal but before receiving his PayPal buyout payout, working on what would eventually become known as the Mars Oasis project?

The goal of the project was to launch a small greenhouse into space and control it remotely until it landed on Mars, ready to operate and filled with everything an astronaut would need to bring a bit of green to the red planet. By doing so, Musk also hoped to rejuvenate interest in space travel to a new generation which he believes is crucial as it is, statistically speaking, just a matter of time before an extinction level event makes earth much less habitable that it currently is.

Starting in 2001, Musk began building a relationship with dealers of ballistic missiles in Russia. His initial welcome was cold, however, as the Russians felt he did not know what he was talking about. During his next trip to Russia in 2002, Musk proved he was now an expert on the topic which is why his new friends offered to sell him one gently used ballistic missile for $8 million. After running the numbers, Musk came to a realization that caught him completely off guard, it would be cheaper for him to start a company and build his own rockets than it would be to buy a single Russian missile.

Once he began to run the numbers even more thoroughly, Musk also realized that the profit margin on space faring vessels was astronomical, with only slightly more than 1 percent of the sale price being enough to cover the entire cost of construction. Plus, with a company whose express purpose is to build and sell rockets, it would be much easier for him to design a ship capable of successfully making it all the way to Mars.

With the numbers supporting the decision, Musk soon formed Space Exploration Technologies (SpaceX) and, working with the team he had assembled, soon developed designs for a vehicle capable of reaching Mars successfully while still being cheaper than any other rocket on the market by an astronomical 90 percent! What's more, the SpaceX design still returns an average of 70 percent profit of the investment cost per rocket.

To get SpaceX up and running, Musk spent roughly $90 million of his own money on the project simply because he thought it was the right thing to do. SpaceX's motto is to make humanity a spacefaring race and they are already doing their part to make this dream a reality.

Prior to coming up with the Mars Oasis project, Musk had never so much as taken a class in jet propulsion, much less rocket science. Unwilling to let something like a lack of any prior knowledge stop him, he became friends with a local rocket scientist and asked to borrow his research on the topic.

When the rocket scientist asked which research, Musk simply gestured to the nearest bookcase full of books on the topic. He then took them home and read them all. When this didn't lead him to the level of knowledge he was hoping for, he then went out and hired every single available rocket scientist and picked their brains until he was comfortable he knew everything there was to know about the topic.

Remember, this was the level of research and dedication that Musk put into the topic when he was still simply planning on buying a Russian missile. Nevertheless, during this phase he was already brainstorming ideas of his own and consulting with his rocket scientists on the plan that would ultimately lead to Falcon 1, the first rocket ship his team built which he named in honor of the Millennium Falcon.

Additional sources of inspiration for SpaceX came from the *Foundation* series written by Isaac Asimov. Musk has frequently stated that Asimov's views regarding the correct application of space travel moving forward is a crucial step when it comes to expanding the human consciousness beyond its currently relatively limited scope. Currently, Musk says that humanity has taken about as long to develop space travel as it did to crawl out of the oceans millions of years ago which means we are just about due for a major upheaval.

Since its inception, SpaceX has generated a pair of unmanned launch vehicles, Falcon 1 and Falcon 9, as well as a fully functioning unmanned spacecraft christened Dragon. Falcon 1 was the first vehicle that had even made it to orbit using liquid fuels and launched by a private company when it had its first flight in 2009. In 2012, NASA signed a contract with SpaceX to put Dragon solely in charge of refueling the international space station as well as delivering the astronauts supplies from the surface. Dragon is now used instead of the actual space shuttle for space flights which means Musk reinvented space travel in less than 20 years.

Not even close to satisfied, SpaceX launched its first satellite into orbit in 2013 and in 2015, the Falcon 9 launched a more advanced satellite built to observe the climate of deep space and determine just how exactly solar flares affect electromagnetic fields on earth. For the next step in this process, SpaceX is currently trying to get permission from the United States government and additional governments worldwide to launch a host of 4,000 satellites which would then be used to ensure everyone had access to reliable and fast internet connections.

Additionally, SpaceX can currently be said to be the most prolific creators of rocket engines in the entire world and their Merlin 1D model engine, which is powerful enough to lift over 40 cars, is currently used for numerous purposes around the world. SpaceX was given its first NASA contract in 2006 along with nearly 2 billion dollars to get the Merlin 1A working properly. These tests ultimately led to the creation of the Falcon and the Dragon.

With SpaceX, Musk hopes to decrease the cost of space travel to the point of making it a realistic notion again. His plan is to send a manned mission to Mars by 2030 and a colony of nearly 100,000 by 2040. He has been quoted as saying, to no one's surprise, if he has anything to say about it, everything on Mars is going to be all electric. To that end, he has also created the Musk Foundation with the hope of determining the best renewable and clean energy sources can be used to make space travel faster, cheaper, safer and more efficient.

Tesla Motors

While Musk was busy worrying about rockets, numerous other engineers were working on the prototype for what would eventually be known as the Tesla Roadster. Having been thinking about electric cars for the better part of 2 decades, however, Musk came on board as soon as the project requested funding. He invested heavily in the project and took on the role of chairman of the board at that time as well. This position also allowed him to take on a more active role in the design of the soon-to-be commercially released Tesla Roadster.

Currently, Tesla Motors offers 3 different models on the road with a fourth version, priced around the cost of any other similar vehicle, announced in March of 2016 which quickly sold through its preorder allotment. While the number of vehicles it has on the road is still relatively small, Tesla Motors is already being compared to the Ford Company. It is also the first successful new automotive company to be founded in America in more than a century.

While he was certainly involved in the coming together of the Roadster's ultimate design, Musk did not begin to take on a more active role of managing the company until 2008, around the start of the Global Financial Crisis. He took on the role of CEO at that time as well as the role of product architect as well. While the product line has since become a success, it was far from a sure thing at several points in the production process.

Musk first met Marc Tarpenning and Martin Eberhard, the creators of the original Tesla model in 2001 when the duo went to listen to Musk speak on his newest passion project of traveling to Mars. They exchanged pleasantries and nothing much came of the encounter except that Musk remembered their names when they met again three years later in 2004 to pitch Musk on the idea for an electric car called TZero.

Musk was instantly hooked on the pitch and arranged to meet with the pair the very next week. The meeting, scheduled for a tight 30 minutes, quickly grew to be more than 3 hours in length as the trio discussed the specifics of the TZero as well as the importance of creating a vehicle that could compete on the road as well as at the pump. This meeting was also the genesis of the rollout strategy that Tesla Motors would ultimately employ to great success, starting with a high-end model to capture the hearts and minds of the public before rolling out a mainstream edition to take advantage of the fervor.

This early brainstorming session also called for the first vehicle in their stable, the Tesla Roadster, to roll of the line, with the assumption that the company would be easily turning a profit by 2008. This ultimately proved to be significantly too optimistic, however, as the coming weeks and months would see Eberhard and Musk butting heads frequently while the creation of the vehicle itself languished amidst numerous production and engineering issues.

It was during this period that Musk also took a more hands on role in the creation of the company's flagship vehicle, making several changes including the door placement and the decision to create unique headlights, setting things back substantially. Further delays accumulated as he saw the need to redesign the seats, the style of the interior and even redesigned the transmission from the ground up. While the changes were made to enhance the overall look, feel and quality of the vehicle to ensure it was worth its premium price, they pushed the already tight production schedule further than it could allow.

The decision to use unique parts also put the fledgling company into a positon of having to source the creation of unique parts, a process no one working at the company at the time was actually familiar with. This lead to Musk shouldering even more responsibility, going so far as to visit Lotus, the manufacturing company in England that would ultimately create each Roadster just to try and get a handle on the spiraling production line. Despite these hardships, Musk held onto his desire for the Roadster to be noticeable, and accessible and it ended up being the primary creative force behind the ultimate result.

In 2007, after fighting with Eberhard nearly every step of the way, and having a particularly heated argument where Musk was credited with the creation of the Roadster in an article and Eberhard was not mentioned, Eberhard left the company amidst a flurry of lawsuits and countersuits by Musk. At this point Musk took on the duties of CEO, making his first act in the positon to fire a quarter of the existing staff as the protracted production schedule meant the company was hemorrhaging money and it was the only way to ensure his $50 million investment wasn't going to become a huge mistake.

In 2008, the first Tesla Roadsters finally rolled off of the line and the reviews were middling at best and terrible at worst. By 2010, 75 percent of the initial run had been recalled due to various hardware and software errors. The initial run of the Roadster ended with 2,150 units being produced and shipped to more than 30 countries worldwide.

Not to be deterred by what he saw as just a wider field test for the concept, Musk doubled down on his Tesla Motors investment and began taking preorders for the follow up to the Roadster called Model S. Following the wave of positive early feedback regarding the Model S, the company filed an IPO worth $100 million, roughly twice as much as Musk had sunk into the company thus far. At the start of 2016, the company was estimated to be worth $25 billion. This still isn't enough for the current CEO who says he expects the company to be worth nearly 30 times as much before 2030.

After the Model S launched to significantly improved reviews and started taking off in the way that Musk and the other creators had envisioned all those years before, Tesla Motors introduced a 4-door variation of the Model S and their first electric sports utility vehicle, the Model X. The company has also begun to manufacture the powertrain system that drives the current electric offerings from Toyota as well as Mercedes.

When he took over as CEO, Musk saw the biggest problem related to the widespread adoption of the electric car to be its limitations when it comes to the ability to travel an extended distance, regardless of the size of an individual battery. With this in mind he began an initiative to increase the number of charging stations available across the United States. In the past 8 years this initiative has more than tripled the availability of charging stations nationwide.

In 1992, a pair of Musk's cousins asked for his thoughts on investment opportunities in California. Musk pointed them towards solar power and in 1993 SolarCity was born. In 2007 it became the largest single supplier of solar power in California. Musk is also chairman of the board on this company and it was a large part of the charging station initiative both in California and across the country.

Musk's goal throughout his tenure as the CEO of Tesla Motors has been to increase the widespread acceptance and usage of electric vehicles in general, not just of those flying the Tesla

Motors flag. To this end, he has also released all of the patents related to electric motor technology that the company previously held. This means that anyone is welcome to use their designs as long as it is done in good faith, with the intent of increasing the overall development of the product space. Currently Musk's salary with the company is just $1 with anything else being generated from stock options and performance bonuses.

Despite having been in the public eye for the better part of a decade, Tesla was only able to acquire the rights to Tesla.com in 2016. Prior to this point, the domain was held by a man named Stuart Grossman who first purchased the domain in 1996 before anyone had thought of buying up obvious domains for future profit. Grossman wasn't using the domain for anything, but he liked the idea of being able to at some point in the future which is why it took a personal visit from Musk to ultimately convince him to sell. While the ultimate price of the transaction was not disclosed, a close friend of Musk's named Jason Calacanis was quoted after the fact as saying it would have been worth several million in Musk's eyes to secure the domain in question.

Chapter 5:
What's Next

As a college student, Musk asked himself what three technologies were going to change the world the most in his lifetime and then went to work on making his mark on all of them. As people are only a few years away from being able to realistically have one of his cars drive them around while they use PayPal to buy something with help from internet from his satellites, it is safe to say he hit his goals. This doesn't mean he is ready to give it all up, however, as he has additional plans that ensure the future will be the best possible version of itself it can be.

Hyperloop

In 2013, Musk went to a presentation of the state of California's proposition for a rail system that would travel at high speeds between Los Angeles and San Francisco. He didn't think the system that many engineers had spent years developing was actually all that fast so he decided to come up with something better. To do so, he got all of the engineers from his electric car company and all of his rocket scientists together and worked with them to come up with what they are calling the Hyperloop. The Hyperloop is a way of traveling long distances at speeds of 700 miles per hour inside capsules that travel on cushions of air.

The finalized design of the Hyperloop, as well as ancillary devices, took roughly a year to create and once it was finished, Musk released the schematics online and declared them open source for everyone to use. He then announced a competition, sponsored by SpaceX, to design the best pod for use in the

Hyperloop system. This competition, currently underway, pits teams from around the world against one another, testing their real world designs on a massive test track that Tesla created for just this purpose. The winners will receive opportunities to further enhance their designs via SpaceX and will be chosen based on the practical application of their designs in late 2016.

Open AI

In 2015 Musk revealed the existence of yet another initiative, this one a nonprofit that is dedicated to researching and creating artificial intelligence in a way that ensures it evolves to be both safe and beneficial for everyone involved. Specifically, Musk has stated that he wants the organization to stand against any potential abuse of the masses that could be perpetrated by an artificial intelligence that is created for such a purpose either by a major corporation or world government.

Musk isn't alone in this endeavor either, and has secured the alliance of other notable scientific minds including Stephen Hawking who believe that artificial intelligence may well pose the most real-world danger when it comes to its potential to negatively impact humanity's survival rates long term. This is why Open AI exists, to ensure that artificial intelligence benefits mankind instead of destroying it. Like most of the technological breakthroughs he has been associated with, all of the work that Open AI does will be open-source and freely available online. Musk is the co-chair of the project and is very aware that the organization needs to tread carefully to ensure they don't accidentally create the thing that they fear the most.

He believes that the best way to prepare against what could be considered the darkest timeline, is to prepare the masses for what might be in the near future. Musk hopes to provide enough readily available information and programing that everyone has the tools to protect themselves if, or when, the time comes.

While Musk believes that the nonprofit will eventually create something that surpasses the intelligence of its creators, he doesn't expect that to happen any time soon, not for 30 years at least. The project is funded and ready for the challenge, however, as individuals from around the world have already pledged over $1 billion to ensure the organization's coffers are full for years to come. Artificial intelligence is the one vision that Musk has not yet realized and this time there is hardly anyone anywhere who is willing to bet against him.

Chapter 6:
Solar City

SolarCity corporation is an American energy services provider, with its headquarters based in California.

Its primary services are the design, sale, financing, and installation of solar and sustainable power systems throughout the United States.

SolarCity is also chaired by none other than Elon Musk. Musk clearly has a large interest in the sustainability of the Earth and its inhabitants, as evidenced from his investing in projects such as Tesla and SpaceX.

SolarCity was originally founded in 2006 by Lyndon Rive and Peter Rive – Elon Musk's cousins. Elon is now the largest shareholder of SolarCity and has even proposed a 2.6-billion-dollar deal that would see his Tesla company purchase SolarCity.

Musk is envisioning the marriage of the two companies, which would see Tesla motor vehicles be fitted with an entirely new roof, completely made up of solar panels.

His vision is to create "the world's only vertically integrated energy company offering end-to-end clean energy products". The customer would be able to have a SolarCity panel installed on their roof, capture the energy from the sun and store it in a Tesla battery, and then use that to power their house or car. This would allow people to essentially live off of the grid with the assistance of just one company!

Whether or not this deal will go through still remains to be seen however, as the decision comes down to Tesla's board members.

There is a bit of hesitation surrounding the deal, as created by shareholder pressures.

Some shareholders and also people in the media are viewing this proposed buyout as a bad deal. SolarCity is currently not profiting, and has experienced a large drop in shares – making its current market value a lot lower than the suggested 2.3 billion-dollar purchase price.

The deal obviously has its risks, but Elon Musk is no stranger to risk and has come close to bankruptcy before with Tesla Motors. However, with big risk comes big reward. If the demand for solar energy continues to rise, over the coming decades Elon Musk, Tesla, and SolarCity together would be in a great position to take the lion's share of the market.

Will the deal come to fruition? Who knows. But either way, Elon Musk is making big moves, and will continue to make them, in the area of solar and sustainable energy production and distribution.

Chapter 7:
Elon's Investments

Over his career, Elon Musk has made many different investments. In this chapter we will share the different companies he has invested in over the years, both the successful ones, and the failures!

Zip2

Musk's very first company was Zip2. We have already discussed this company in an earlier chapter, but essentially Zip2 was the internet's first yellow pages, and could be considered quite a big success. Musk sold the company in 1999 to Compaq for $307 million. At the time, that was the largest amount ever paid for an online company.

X.Com

Also previously discussed, Musk took $10 million from the sale of Zip2 and used it to start X.com, an online payment service. Eventually, X.Com merged with a competitor called Confinity, which later changed names to Paypal. Musk had some issues with his new business partners and was eventually voted off of the board.

It wasn't all bad news for Elon however, as Paypal was acquired by eBay in 2002 for $1.5 billion. Musk was not a majority owner by any means, but he still netted over $100 million from the sale – a great return on his initial $10 million investment.

Everdream Corporation

Elon Musk's cousin, Lyndon Rive was a co-founder of the company Everdream. Everdream sold desktop management services to small businesses, and performed tasks such as fixing antivirus software, and backing up data.

Musk invested in Everdream in the 4th round while he was still involved with Paypal. Almost a decade later, Everdream was sold to Dell in 2007.

SpaceX

In 2002, Elon Musk founded Space Exploration Technologies, otherwise known as SpaceX.

Elon believes that this project is one of the most important things to happen in the history of the Earth. Obviously, it is his most prized company and the one which means the most.

The goals of SpaceX are to make rockets more affordable, and also to allow humans to become an inter-planetary species.

SpaceX has had it's share of doubters, naysayers, and financial issues. However, at this current moment in time the future looks quite bright for SpaceX! In April 2016 SpaceX made huge ground by launching and successfully re-landing a Falcon rocket.

In the coming years, Musk aims to send humans in his rockets to Mars, to begin colonization of the planet.

The Musk Foundation

In 2002 Elon founded the Musk foundation alongside his brother Kimbal. The foundation awards grants that support research into sustainable and renewable energy, space exploration, education, and childhood diseases and disorders.

Tesla Motors

Musk invested in Tesla Motors back in 2004.

The company was first founded in 2003 by Martin Eberhard and Marc Tarpenning. After his investment, Musk joined the board of directors.

Since then, Elon has been heavily involved with Tesla Motors with the aim of making electronic cars the way of the future.

After the market crash in 2008, Elon became the CEO of Tesla, a position which he still holds today.

Tesla has had several financial problems over the years, and has required an enormous amount of investment. However, the company recently broke records by pre-selling its model S sedan, totaling over 325,000 sales in a single week!

The model S sedan starts at $35,000 and is expected to begin production at the end of 2017.

Surrey Satellite Technology

In 2005 Musk purchased a small 10% stake in Surrey Satellite technology, a small-satellite provider.

The idea behind the investment was not so much focused on making a return, but rather for SpaceX to gain a better understanding of how they could one day work alongside this company in providing small and inexpensive spacecraft.

SolarCity

Previously mentioned in this book, Musk has invested into SolarCity.

Obviously being very passionate about sustainable and green technology, this company is an obvious investment for Elon. Currently, Elon Musk is acting as the chairman of the company.

Mahalo.com

Mahalo.com is a question and answer site that was first founded in 2007. The site essentially allows people to ask and answer questions.

However, when Google changed it's algorithm in 2011, Mahalo experienced a large drop in business and was forced to lay off 10% of its staff.

The company has since changed its strategy, and now primarily focuses on how-to videos and live question and answer sessions.

Stripe

Stripe launched in 2010 and is Paypals main competitor. After his experience with Paypal and X.com, Elon Musk saw this as an abvious investment to make.

The company currently provides payment processing services to online apps and companies such as Lyft, Facebook, and Twitter.

Stripe has experienced rapid growth and was recently valued at $5 billion. This one was definitely a good investment for Elon Musk.

Halcyon Molecular

Halcyon Molecular was founded in 2008, with the goal of unlocking the biggest secrets hidden in DNA.

The company had large goals, and aimed to provide a human genome sequencing service that cost less than $100.

However, due to stiff competition it wasn't meant to be. In 2012 the company closed its doors, becoming one of Elon's first failed investments.

Tesla Science Center

Not so much an investment as it was a donation, Musk gave $1 million in funding for the development of a new science museum. The museum is named the Tesla Science Center and is located in New York.

The museum, named after the great scientist and inventor Nikola Tesla, was created in 2014.

Elon Musk plans to build the world's quickest Tesla charging station in the museum's parking lot!

Vicarious

While Musk is not a fan of artificial intelligence, he invested in AI company Vicarious in 2014.

The company has been unusually secretive with its projects, but has stated that their goal is to create a computer that thinks like a human, but does not need to sleep or eat.

Elon reasoning behind investing in AI companies is to ensure that AI is moving in the right direction, and is not being used or created for destructive purposes.

DeepMind Technologies

DeepMind Technologies is another AI company that Elon Musk has invested in to. The company was initially founded in 2011, but was acquired by Google in 2014, changing its name to 'Google DeepMind'.

The company aims to combine machine learning and neuroscience into a powerful, all purpose, computer algorithm.

Future of Life Institute

Once again, more of a donation than an investment, Elon Musk gave $10 million to the Future of Life Institute in 2015.

The institute supports research into the risks associated with developing technologies such as AI. This further demonstrates how worried Elon Musk is about the possibility of artificial intelligence being used for negative purposes.

NeuroVigil

NeuroVigil is a startup that was launched back in 2007. The company developed the worl'd first portable brain monitor, the iBrain.

Musk became a principal investor in the company in 2015.

The company uses its technology to help drug companies conduct clinical trials, as well as to diagnose and treat patients with neurological disease.

NeuroVigil also wants to help NASA keep track of their astronaut's brains whilst they're aboard the International Space Station. This could be a big reason for Elon's interest in the company.

Hyperloop

Though he hasn't physically invested in the business yet, Hyperloop looks to be Musk's next big project.

He first announced the idea for Hyperloop back in 2013, through which he wants to build a series of high speed rail systems that would transport people at speeds exceeding 500mph!

There is not much information about the proposed Hyperloop at the current time, but stay tuned – the possibilities for this technology look to be huge!

OpenAI

Another artificial intelligence company, OpenAI, was first founded in 2015.

However, this time, Musk was one of the co-founders.

This companies primary focus is to become a repository of research papers, blog posts, code, and patents that leading scientists can contribute. OpenAI aims to allow people to safely and openly work on AI.

Elon Musk recently contributed the the $1 billion dollars of funding that the company has raised so far.

Chapter 8:
Personal Life

Elon Musk has also had a pretty interesting personal life throughout the years.

The pressure of being constantly in the media, alongside with financial struggles and stress from work, can put a lot of strain on a person's personal life at home.

Reportedly working 100+ hours weeks, it is easy to understand how the man's family life can take a hit.

Elon was initially married to Justine Wilson for 6 years, a relationship that produced 5 sons. Shortly after their divorce in 2008, he met Talulah Riley in a London nightclub.

Musk was at the club, trying to clear his head after the recent divorce. Introduced by the club's promoter, Elon was instantly impressed by the actress, and thus began their relationship.

Elon was impressed by Talulah's intelligence and interest in talking about rockets and electric cars. He quickly realized that she was not simply just another model, but had intellect to match her looks.

They met up again several weeks later in Beverly Hills, where lying in bed, Elon spontaneously asked her to marry him. He didn't have an engagement ring, but later bought her 3: a giant one, and everyday ring, and another with a diamond surrounded by 10 sapphires.

At the time however, his divorce was not settled with his first wife. He was also experiencing financial troubles with SpaceX and Tesla, and it was a very stressful time for the new couple.

Talulah described him during this period by saying 'He looked like death itself'.

Amongst all of the stress, the couple divorced, re-married less than a year later, and almost divorced for a second time. They changed their minds and Musk tore up the divorce papers before their second divorce was processed.

Elon appears to be a tough man to live with, going through almost 3 divorces in a very short period of time. His first wife, Justine described him by saying "Elon does what he wants and he is relentless about it. It's Elon's world and the rest of us live in it".

A determined and focused man, Elon Musk is the product of a tough upbringing in South Africa, where it is said he experienced abuse from his father.

A victim of severe bullying growing up, Elon appears to be left with permanent emotional scars. Justine referred to him as being "a tortured soul".

Whilst his rough up-bringing has molded him into a sometimes difficult man to live with, it appears to have blessed him with tenacity and focus. He has a level of relentlessness that very few others have, and perhaps that is what allows him to be the incredible entrepreneur that we know today.

Chapter 9:
Artificial Intelligence

As you might have noticed from Elon Musk's different investments, he holds a great deal of interest in the area of artificial intelligence.

This interest however is coming mostly from a place of fear and worry about what the future of artificial intelligence may hold. One of Musk's most recent investments, OpenAI, is focused on making the future of artificial intelligence safe and sustainable.

There is a very real possibility that artificial intelligence could soon surpass the intelligence and capabilities of humans. There are several movies depicting this possible demise, such as iRobot and Terminator, but a version of these fictional shows could possibly eventuate.

Elon Musk has stated that for the most-part he isn't too worried about artificial intelligence. He has investing in many different AI companies just to keep his finger on the pulse of what's going on in that industry. However, in a recent interview he said there is really only one company that he's worried about. He was reluctant to say which company he was talking about, but strongly hinted towards Google.

The other main threat that Elon has mentioned in the area of AI is something called the Cyber Grand Challenge. The Cyber Grand Challenge is basically a hacking tournament held in Vegas.

The Challenge is run by the Defence Advanced Research Projects Agency, also referred to as DARPA. DARPA states that their goal is "to find new strategies for countering cyber-warfare".

Musk however isn't a big believer in the validity of this mission. He fears that the hacking challenge may result eventually in creating something similar to the Skynet AI that existed in the Terminator movies. Basically, he's worried about it leading to the creation of an all-powerful supercomputer.

DARPA seeks to create an automated artificial intelligence system that can detect and resolve bugs in a computer system. Leaving a computer up to this task without the assistance and decision making powers of an actual human could potentially be disastrous.

For most people, the possibility of a malicious artificial intelligence system seems hundreds of years away. However, with the current rate of technological development, it could occur a lot sooner than many would expect. Luckily, Elon Musk is very aware of this issue and is using his power and influence to raise awareness.

The future of artificial intelligence is for the most part extremely exciting and presents some amazing opportunities for man-kind.

If it can be developed in a safe way, with Elon Musk's involvement of course, then this is an area that we should have nothing to worry about!

Chapter 10:
What Will Be Elon Musk's Legacy?

So, at the end of the day, what will Elon Musk be best known for?

What will be this incredible man's legacy?

Will it be his work in the area of artificial intelligence? Will it be for his contributions to sustainable energy through his Tesla cars, and work with SolarCity? Or perhaps it will be his role in potentially populating Mars!

Elon Musk has a range of projects in the works, from attempting to populate Mars, to creating sustainable energy for the world, to creating a Hyperloop system that could change the way people travel!

With so many different projects, it's hard to know exactly what Musk will be be known for. The billionaire entrepreneur is still only in his 40's. He has decades and decades of potential ahead of him, so for all we know he will begin a new project altogether that will be the one we remember him most for.

But, in this author's opinion, Elon Musk will be best known as an inventor of sorts. A powerful entrepreneur who's work has literally changed the face of the world. Similar to people such as Albert Einstein, Thomas Edison, Nikola Tesla, his work will be remembered worldwide for years and years after his life his over.

His legacy will include a range of inventions, projects, philanthropic efforts, and inspirational moments where he was able to overcome the odds stacked against him.

For anyone who has big aspirations as an entrepreneur, inventor, or in just about any worthy endeavor, Elon Musk currently is and will remain to be an incredible source of motivation and inspiration that it can be possible. Being an inspiration for future generations may very well be the most valuable part of his legacy when it is all said and done.

Chapter 11:
Lessons to Learn

If you take a long hard look at the lives of great individuals who leave an indelible mark on history, you can always find a few takeaways or words of wisdom to live by. It doesn't take much effort to find lessons worth emulating in the life of Elon Musk, and putting them to work on a personal level can make it easier to believe in yourself and to follow your dreams both in the short and the long term.

Be aware of the signs around you and act when appropriate: Musk isn't a successful force in several exceedingly advanced and profitable industries because he is extra smart, rich or even extremely lucky - though he is all 3. He is successful because when he was in college he looked around the world as it was at the time and found the indicators that pointed him in the right direction, first to the internet, then to space and electric cars. Anyone in a similar situation at the same time could have seen the same thing but Musk was the one who saw the signs and took the time to interpret them and put their information to good use.

Musk saw what needed to be done, and then did everything he needed to in order to ensure that he was in the right place so that when the right time came along he was ready for it. What's more, he didn't let the fact that following through was difficult, dissuade him from taking advantage of the success he knew was on the way. Being insightful is useful, as is being hardworking or dedicated to a cause; nevertheless, being all three and understanding how to best apply each skill is what really makes a person successful.

Take everything in stride: When thinking about the success that Musk has had over the years, it is equally important to remember the adversity he faced first when he was nearly beaten to death, again when he was ousted from Zip2, again when he was ousted from X.com and finally the drama prior to the successful launch of the Tesla Model S. In each and every one of these situations he could have turned his back on the hard road and found something easier without a second thought.

Musk could have been home schooled and then stopped working for life after either of his internet companies were ultimately sold off and finally, he could have left Tesla Motors as a grand failed experiment after the Roadster and focused on his completely successful business of building rockets. Instead, he used each setback to push himself forward to new and previously unprecedented heights. The lesson to learn from this is clear, never let events that appear terrible at face value get you down, take each as a call to action to improve yourself in new and fantastic ways.

Never stop innovating: From the moment he thought tocreate the online map database with local business information, Musk has been innovating. At X.com he could have been happy focusing on person to person email transactions and never gone on to help create PayPal in its current form, or he could have simply bought a rocket from Russia and never formed SpaceX. He could have remained a simple investor at Tesla or thought the proposed California High Speed Transit plan was fast enough. Instead, he saw flaws in systems that other people would have considered good enough and instead of changing how he looked at these problems he decided to change the world to suit his needs.

The lessons to be learned from this are obvious, if you can't find the right niche for yourself, create something new. If you never stop innovating on your past successes, you will never stop being successful. Follow the more difficult path and if you persevere you will find success.

Work harder than the competition: Musk is currently the acting CEO of a pair of companies that are each worth more than $1 billion and he has a reputation of being into micromanaging. He is also chairman or co-chair of numerous other boards, head of numerous charities and is currently judging the Hyperloop competition. Musk is known to work 100 hours per week because, in his words, if you work more than twice as hard as everyone else you will get 3 times as much done each year. Hard work and perseverance are the true backbone of every type of success.

Seek out your purpose: At age 18 Musk was already talking about electric cars and at 22 he was looking to the stars. SpaceX was formed with the intention of making humanity into a spacefaring race by reigniting the world's interest in space travel. Musk doesn't have small dreams or take partial action, he is all in because he believes he is following the path he was meant to follow. If you ever hope to achieve even a tenth of what he has accomplished, you need to take the time to have an in depth conversation with yourself and determine if you are really doing that which you were meant to do. Once you find your true purpose, the next step is to stop at nothing until you have achieved it.

Be flexible: While you should have a plan to achieve whatever it is your true purpose turns out to be, it is important to not pursue it if all of the external evidence points in the opposite direction. This is how Musk was able to respond to the information that the missile he wanted to buy from the

Russians was way out of his price range and successfully create SpaceX in the process. He was also forced to pivot when the Tesla Roadster's production schedule had spun wildly out of control to the point that he was in a positon to almost lose his investment. Instead he rolled with the punches, did what needed to be done, no matter how difficult it might have been, and persevered until he found success.

Determine your own level of success: When Zip2 was sold for $300 million, it was the largest deal of its kind that had ever been completed, a metric of success in almost anyone's book. Not for Musk, however, as he was convinced that the service could have been so much more than what it ultimately became. He saw failure where many people would have been happy to find success and it drove him to even greater heights because of it. Like Musk, it is important to listen to yourself and not let other's visions of success or failure color your own perceptions of your actions.

Conclusion

Thanks for making to the end of the book! Hopefully, learning from the life of one of the greatest minds of a generation has provided you with numerous insights into how you can improve your own life by following through on the things that made Elon Musk a true success. If imitation is the purest type of flattery, then you can certainly do worse than flattering the modern Henry Ford.

The next step is to stop reading and to start putting the lessons you have learned in the preceding pages to work in your own life. Remember, as long as you take advantage of the signs the world around you is presenting you with you really can't go wrong. Follow your own path, commit to your own success and accomplish anything you set your mind to.

Finally, if you enjoyed this book, please leave a review on Amazon and let others know what you think. It'd be greatly appreciated!

Made in the USA
Middletown, DE
18 December 2016